Food Chains

Theresa Greenaway

HODDER
Wayland
an imprint of Hodder
Children's Books

CYCLES IN NATURE

Other titles in this series:
Plant Life
The Water Cycle

All Wayland books encourage children to read and help them improve their literacy.

✓ The contents page, page numbers, headings and index help locate specific pieces of information.

✓ The glossary reinforces alphabetic knowledge and extends vocabulary.

✓ The further information section suggests other books and websites dealing with the same subject.

Cover: A grizzly bear eating a fish (*main image*); (*clockwise from top right*) a colourful variety of vegetables; a diagram of a food chain; ears of ripening wheat; a cow grazing on grass.
Title page: A grizzly bear eating a fish.
Contents page: A colourful variety of vegetables.

Series editor: Nicola Wright
Book editor: Alison Cooper
Series and cover design: Sterling Associates
Book design: Jean Wheeler

First published in 2000 by
Hodder Wayland a division of Hodder Children's Books
388 Euston Road
London NW1 3BH

© Copyright 2000 Hodder Wayland

Typeset by Jean Wheeler
Printed and bound in Italy byEurografica S.p.a.

British Library Cataloguing in Publication Data

Greenaway, Theresa
    Food chains. – (Cycles in nature)
    1. Food chains (Ecology) – Juvenile literature
    I. Title
    577.1.'6
ISBN 0 7502 2515 7

Picture acknowledgements
The publishers would like to thank the following for allowing their images to be used in this book: Bruce Coleman 18 (*centre*)/M.P.L. Fogden, 23 (*top*)/Andy Purcell, 23 (*bottom*)/Kim Taylor, 24/Christer Fredriksson; Greg Evans 17; NHPA 16/Bryan & Cherry Alexander, 18 (*bottom*)/G.I. Bernard, 19/B. Jones & M. Shimlock, 27/Yves Lanceau; OSF *cover* (*bottom left*)/Niall Benvie, 12/Daniel J. Cox, 20/Mike Hill, 25/Mark Hamblin, 29/Bob Gibbons; Papilio 7 (*top*); RSPCA *cover* (*top left*)/Angela Hampton, 26/E.A. Jones; Still Pictures 4/Jean F. Stoick; Stock Market 5/Daniel Aubry; Tony Stone Images *cover* (*main image*) & *title page*/James Balog, 8/Nicholas Parfitt, 9/Daniel J. Cox, 11/Mark Petersen, 21/Kathy Bushue, 28/Darryl Tucker; Wayland Picture Library *cover* (*top right*), *contents page* & 7 (*bottom*). Artwork on pages 6, 10, 13 and 14–15 is by Peter Bull. Artwork on page 22 is from Wayland Picture Library. Border and activity box artwork is by Jan Sterling.

Contents

Eating to Live

Every living thing is eaten by something else.
Dead or alive, plants and animals are all food for
something. Even wood, bones and fallen autumn
leaves are eaten. The different kinds of living
organism in the wild are all dependent on each
other. A hare may nibble on some leaves. Later,
a lynx may catch and eat the hare. Leaves which
are eaten by a hare, which is then eaten by a lynx,
is an example of a food chain. The lynx is
dependent on the leaves, even though it does
not eat leaves itself.

▼ A hungry lynx
chases a hare.

How many habitats?

Within a large habitat, such as a park, there are many smaller habitats, such as leafy bushes or short grass. How many different habitats can you find in your school grounds or nearby park? What kinds of insect or other invertebrate can you find in them? What kinds of bird use them? Do some birds prefer particular trees or bushes?

All living things, including people, need food. They also need oxygen to breathe, water, warmth and shelter. Each living thing finds these vital resources in its habitat – the environment in which it lives.

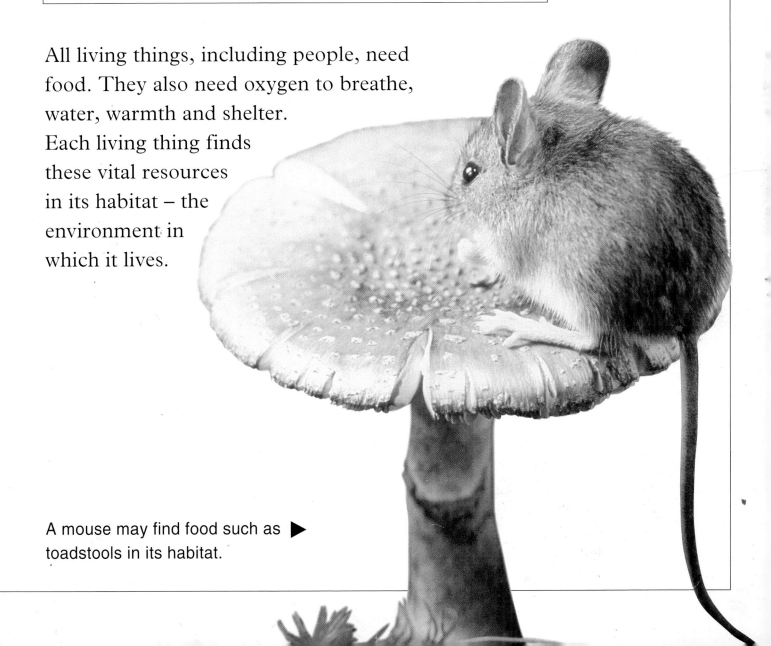

A mouse may find food such as ▶ toadstools in its habitat.

Plant food

Plants are green because they contain a green pigment called chlorophyll. This has the ability to trap the energy in sunlight. Plants use this energy in an important set of chemical reactions called photosynthesis. During photosynthesis, carbon dioxide from the air is combined with water inside the leaves to make sugar and oxygen.

Plants use sugar, and the minerals absorbed from the soil through their roots, to make all the other substances they need to grow.

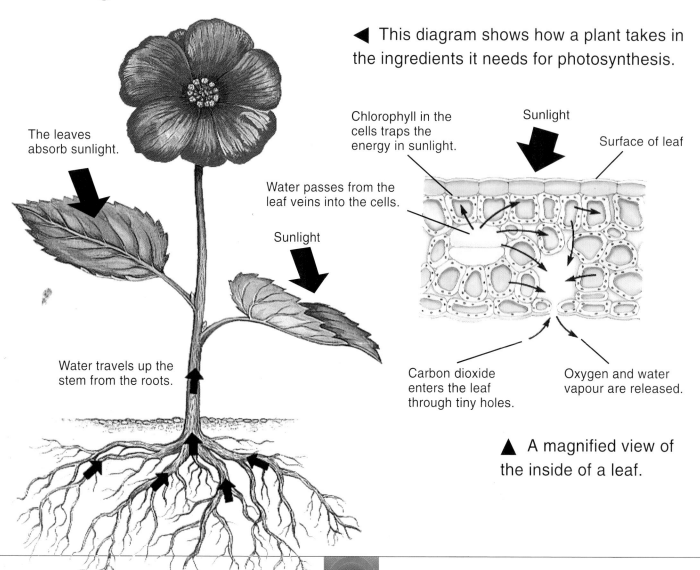

◀ This diagram shows how a plant takes in the ingredients it needs for photosynthesis.

The leaves absorb sunlight.

Chlorophyll in the cells traps the energy in sunlight.

Sunlight

Surface of leaf

Water passes from the leaf veins into the cells.

Sunlight

Water travels up the stem from the roots.

Carbon dioxide enters the leaf through tiny holes.

Oxygen and water vapour are released.

▲ A magnified view of the inside of a leaf.

▲ Even water plants, like these water lilies, rely on the sun's energy to make food.

On land, mosses, liverworts, ferns, conifers and flowering plants grow wherever they can find enough warmth and moisture. Seaweeds and tiny chlorophyll-containing organisms float in the seas. All these photosynthesize, trapping the sun's energy and using it to grow and reproduce. Some energy is stored as starch in swollen roots or stems. Wherever they grow, plants are very important. We could not survive without them.

▼ Vegetables like these provide us with many of the nutrients we need for healthy growth.

Animal food

An animal could sit in the sun all day, but it would not make any food for itself. It cannot turn carbon dioxide and water into sugar – only plants can do this. So all animals depend on plants.

Some animals eat only plants. They are called herbivores. Others only eat other animals. These are carnivores. Some, such as humans eat both plants and animals, and are called omnivores. Scavengers eat the rotting remains of animals left behind by other carnivores, or the bodies of animals that have died naturally.

▼ Scavenging birds called vultures feed on a dead zebra in Tanzania.

Very few animals eat just one kind of plant or prey. If the supply of that one kind of food ran out, they would starve. To ensure that they always have enough to eat, carnivores eat many kinds of other animal and herbivores munch a wide variety of leaves, fruits or seeds.

▲ A tasty trout makes a good meal but otters also eat crabs, frogs, eels and even young birds.

Herbivore, carnivore or omnivore?

Make a list of all the pets that you and your classmates own and work out whether they are herbivores, carnivores or omnivores. Keep a diary of the birds that you see in your school grounds. What do they like to eat? Try putting out separate heaps of food such as crumbs, birdseed, grated cheese or peanuts. Which food does each type of bird prefer?

Energy and Food Chains

Life on earth depends on energy from the sun. Because plants trap and store the sun's energy, they are vital to all other forms of life. Plants are called producers. Animals that get their energy from eating plants are called primary consumers. Animals that get their energy from plant-eating animals are secondary consumers – they receive the energy trapped by plants secondhand.

Producers and ▶ consumers can be arranged in a pyramid diagram like this one.

Top predator

Secondary consumers

Primary consumers

Producers

Top predators

At the top of the pyramid of producers and consumers are large carnivores called top predators. Lions, tigers, crocodiles, eagles and large sharks are all top predators. There are always fewer top predators than the animals on which they prey. If the numbers of prey fall, top predators find it hard to get enough to eat, and so their numbers fall as well.

▼ Lionesses with their prey, a zebra, in Kenya.

Food chains

When food is eaten, some of the energy it contains is transferred to the animal that eats it. The rest is lost as heat that escapes into the environment. The transfer of energy from plants to the animals that eat them, and their predators, is called a food chain. A food chain is a simple way of showing how food energy goes from plant to predator.

Eaten by a fox (secondary consumer)

↑

Eaten by a rabbit (primary consumer)

↑

Grass (producer)

◄ This diagram shows a short food chain. There are three stages in this chain.

▼ This young fox has caught a ground squirrel.

Eaten by a
sparrowhawk

Eaten by a willow tit

Eaten by aphids

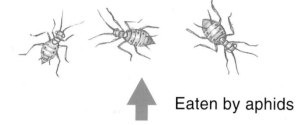

Leaves

How many food chains?

What types of plant and animal live in your garden or school grounds? Look out for insects feeding on leaves or flowers, or caught in a spider's web. There might be birds that eat seeds and fruits as well as birds that eat insects. You might spot frogs, mice and perhaps even hedgehogs. Cats, dogs and foxes might be occasional visitors.

Can you work out how all these plants and animals fit into food chains? How many food chains can you think of? Which is the longest chain?

◀ This diagram shows a longer food chain: it has four stages. Sparrowhawks never eat leaves, but they are just as dependent on them as the aphids are.

Food webs

Usually, many kinds of plant and animal live close together, and most animals eat all sorts of different things. This is why a lot of food chains are connected to each other. This far more complex network is called a food web. Working out a food web helps us to understand how the different kinds of animal depend upon each other, and on the plants that grow around them.

▼ This food web shows how plants and animals in an African grassland depend on each other. The arrows point from the food to the animal that eats it.

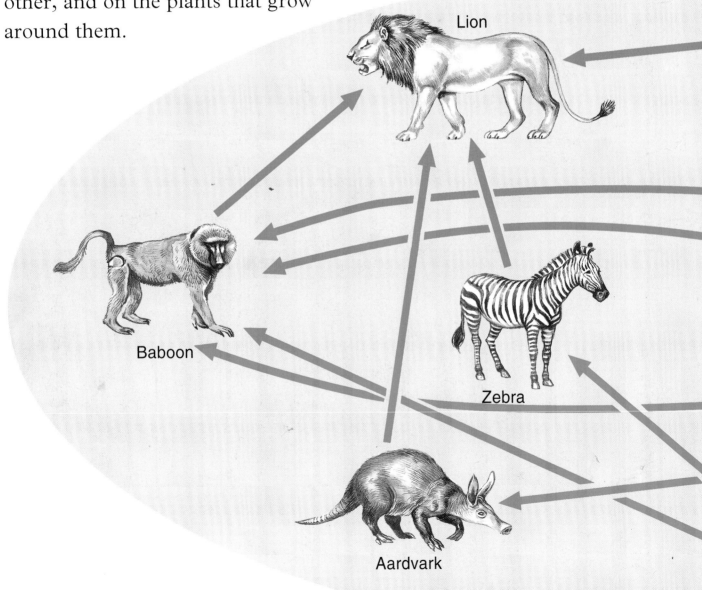

Lion

Baboon

Zebra

Aardvark

Making a food web

Stand in a circle with all your classmates. One of you names a plant, takes a ball of string and holds the end of it. Then someone names an insect that might eat the plant. He or she takes the ball of string and holds on. Someone else names an animal that might eat the insect, and so on. When someone names a top predator, cut the string and start with a plant again. Can you name more than one creature that might feed on it? How complicated a food web can you make?

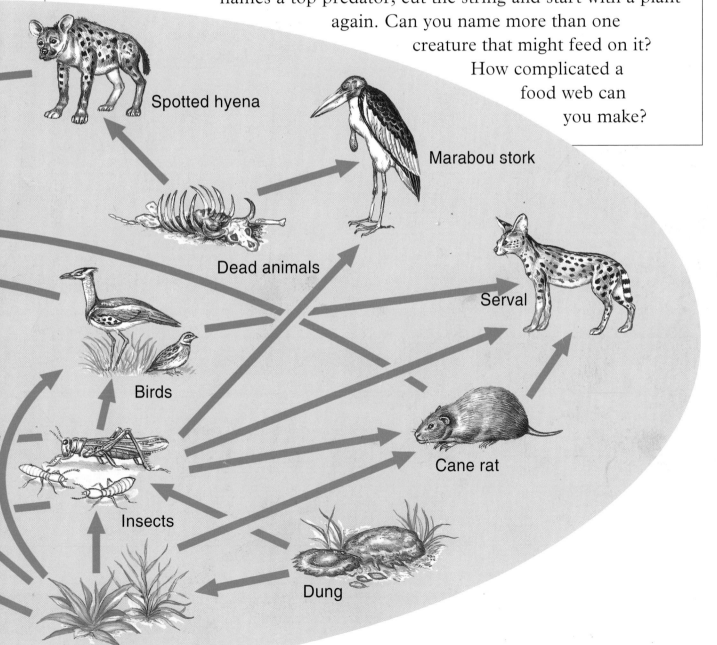

Spotted hyena

Marabou stork

Dead animals

Serval

Birds

Cane rat

Insects

Dung

Human food webs

For hundreds of years, people ate what they could grow or catch in the area where they lived. For example, an Inuit person's traditional food chain would have looked like the one on the right.

As traders started to carry spices, sugar and other products to distant lands, diets began to change. Today, fast air transport means that people can eat foods from many different parts of the world, at any time of year, as long as they have the money to buy it.

Polar bear (possibly)

Inuit

Seal

Fish

Photosynthesizing plankton

▼ An Inuit hunter skins a seal. Seals are the Inuit's traditional food.

But there are still differences between the diets of different peoples. The food webs of people living in rural parts of developing countries are mainly made up of foods they can catch or grow themselves. The people of wealthier countries eat much more processed food. Some foods and drinks seem to be part of the food web almost everywhere. In cities all round the world, you can buy cola, coffee, pizzas and beefburgers.

Are you a top predator?

Try to make your own food web. Write down everything that you eat in one day, and all the other animals that might also eat these foods if they had the chance. Work out your place in the food web. Do you think you are a top predator?

▼ This girl's food web includes foods prepared in processing plants and bought from a supermarket.

Productive Habitats

Because plants are the producers, places where lots of plants grow produce the most food and store the most energy. They are known as productive habitats. They can support large numbers of consumers.

Tropical rain forests grow near the Equator. Here, it is warm all year round and rain falls almost every day. These forests are always green and leafy. The tall trees are festooned with climbers and other smaller plants perch upon their branches. The treetops teem with life – insects, birds, snakes and monkeys. The food web here is very complicated.

▼ Brightly coloured macaws (top) and snakes (bottom) make their homes in the rain forests of South America.

In the shallow, sunlit waters off many tropical shores, coral animals grow into large reefs. Tiny green algae photosynthesize and share their sugars with the coral. The reefs are home to brightly coloured tropical fish, shellfish, sea urchins, crabs and octopuses. Reef sharks cruise over them, hunting for prey. Eels lurk in caves in the coral.

There are millions of different kinds of plant, animal and other living organism. This huge variety is called biodiversity. The greater the number of different living organisms found in an area, the greater its biodiversity.

▲ Shoals of fish dart through the coral reefs off Indonesia.

Less Productive Habitats

In parts of the world where it is very cold or very dry, plants may be few and far between. Little of the sun's energy is trapped and stored. Areas such as deserts, high mountain slopes and Arctic tundra are some of the least productive habitats in the world. They support far fewer consumers than forests, grasslands and wetlands.

▼ Mountain goats have to scramble over rocks to find enough plants to feed on.

▲ Caribou living in the tundra move south during the winter to find plants to eat.

Inside the Arctic Circle, summers are short and winters are very long, cold and dark. Even in summer, only the top few centimetres of the soil thaw. The rest stays frozen solid. This cold land is called tundra. The plants here grow slowly, making a mat over the ground. It is too cold for trees to grow. Arctic animals have to be tough to survive.

How many plants?

Compare the numbers of plants growing in different places in a garden or your school grounds. Are there as many growing on a weedy path as there are in a lawn or vegetable patch? Can you think of a way to measure how productive each of these different mini-environments is?

Recyclers

When a plant or animal dies, the energy it has stored is not lost. Whole armies of scavengers, fungi and bacteria feed on dead organisms and break down their bodies. These recyclers are called decomposers.

▼ This diagram shows how dead organisms are broken down and recycled.

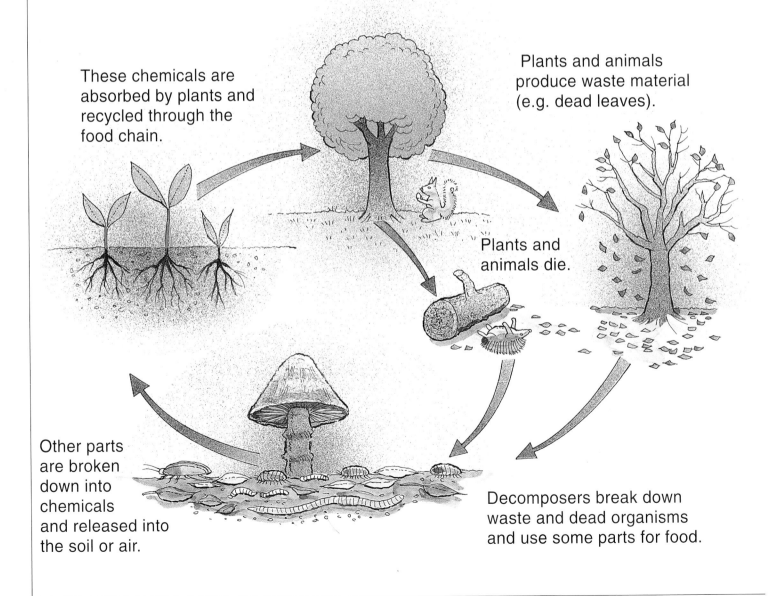

These chemicals are absorbed by plants and recycled through the food chain.

Plants and animals produce waste material (e.g. dead leaves).

Plants and animals die.

Other parts are broken down into chemicals and released into the soil or air.

Decomposers break down waste and dead organisms and use some parts for food.

Slugs, millipedes and woodlice all chew up bits of leaves, stems and rotting wood. They produce large, soft droppings that other decomposers feed on. Burying beetles tunnel beneath small dead animals until the bodies disappear below the surface of the soil. Then they lay eggs in the bodies. Earthworms feed on dead leaves, and also on tiny fungi, bacteria and other remains that are in the soil that they swallow.

▲ A burying beetle gets to work on a dead mouse.

Decomposers at work

You can watch decomposers at work by putting a piece of bread, an apple quarter, half a lemon and a carrot into separate glass jars. Screw the lids on. Keep a diary to record what you see. As fluffy moulds and patches of bacteria grow, the shape of the food changes. Why do you think this is?

◀ Fluffy mould growing on a tomato.

23

Human Interference

Today there are six billion people in the world. People who study human populations predict that by the year 2050 there may be nearly double this number. We will need to grow more and more food, so there will be even less space for wildlife. Food chains and food webs are already being changed and broken.

◀ Tigers are an endangered species but they are still killed in places where people think they may be a danger to their farm animals.

Pesticides

To protect crops from insect pests and diseases, farmers spray them with chemicals called pesticides. Contaminated insects and plants are then eaten by birds and other small animals. The amount of chemical taken in by these small creatures may not be enough to kill them. But a larger predator may eat several of the small creatures and receive a much bigger total dose. This may be enough to kill it, or prevent it from breeding successfully.

▲ Peregrine falcons were almost wiped out as a result of pesticide poisoning.

Raising animals

In the wild, chickens, pigs and cattle find their own food. But farmers need their livestock to grow faster and bigger than wild animals, so they give them extra food. Many people are concerned that some food supplements can damage animal and human health. In the 1980s and 1990s some British cattle developed a disease, called bovine spongiform encephalopathy (BSE), after being given food made from dead sheep, which cattle would not normally eat. Some scientists think people may have developed a similar disease after eating infected beef.

▼ These cattle are eating silage, a food supplement made from grass. Cattle need extra food like this in the winter.

Organic farming

Organic farmers do not use chemicals to kill insect pests and weeds, and they do not give their animals chemicals to make them grow faster. Organic farming methods can help to restore food chains by encouraging predators that control pests naturally. Many people think organic foods are healthier too, because there is no risk of chemicals being passed from the foods to us.

Natural predators such as ▶ ladybirds feed on aphids that attack crops.

Food from the wild

Although much of our food comes from farms, some still comes from the wild. Sometimes we take too much of this 'wild' food, and do not leave enough for the natural predators. People have taken so many cod, haddock and sand eels from the sea that stocks are getting low. Top predators such as seals are often blamed for reducing the fishermen's catches, but really it is people who have ransacked the sea.

Breaking the chains

Play the food web game described on page 15 again. When you have built up a web, imagine that the plant or animal one of you has named has become extinct. That person lets go of the web. What happens to the strength of the web when one organism is lost? What happens to the organisms further along the chain?

Introducing new species

As people travelled around the world, many plants and animals travelled with them. Some, such as pigs, were transported deliberately, to provide food for people in the lands to which they were travelling. Plant seeds often became caught in cargoes. Animals such as rats were unwelcome companions that infested ships.

These 'foreign' species were often released into the wild, either accidentally or on purpose. Often, they had no natural predators in their new land. Many of them have multiplied out of control and caused serious problems.

▼ Pigs damage the plant life of some Pacific islands as they search for food.

On remote islands, such as the Kermadec Islands and the Chatham Islands in the Pacific Ocean, the natural food chains were broken when people arrived. They brought with them pigs, sheep, rabbits and goats that ate too many island plants. Predators such as rats and stoats have disrupted food chains in New Zealand by taking the eggs and young of flightless birds such as the takahe. This bird is now almost extinct.

▲ Rhododendrons brought to Britain have spread through woodlands and prevented other plants from growing.

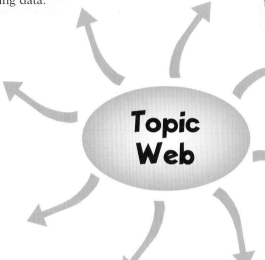

Maths
Multiplication in context.
Venn diagrams and related
   ways of presenting data.

ICT
Practising cut/paste functions.
Using a simple data-handling
   program.

Science
Food chains and food
   webs.
Micro-organisms and
   decomposition.
Habitats: how plants and
   animals are suited to
   their environments.
Photosynthesis in green
   plants.
Using keys to identify
   species.
Using scientific
   vocabulary.

Literacy
Identifying scientific and
   technical language.
Writing explanatory text.
Writing a simple narrative.

Geography
Knowledge of locations.
Humans as agents of
   environmental change.

More Activities

Literacy
• Look through the book for important scientific words.
  Put these words into your own 'Food Chain Dictionary'.
  Remember to put the words in alphabetical order. Then
  write an explanation for each word.

• Choose one of the food chains shown in the book and
  use it as the plot for a story. You could start like this, for
  example: 'Once upon a time a green aphid was feeding
  on a leaf …'. Try making your story into a picture book.

Science
• Make a list of all the animals in a mini-environment.
  Draw a picture of each creature. Then make a key to
  identify each creature. Remember, each branch of the
  key must start with a question that has a yes/no answer.
  You could try making a key for all the plants too.

Maths
• Find out more about the foods that different animals
  eat. You could investigate the foods eaten by animals
  that live in a wood or in grassland, for example, and
  decide whether they are herbivores, carnivores or
  omnivores. Put the information into a Venn diagram.

• Make a list of what you eat in one day. How many
  potatoes? How many slices of bread? How many grams
  of chocolate? What multiplication calculations do you
  need to do to find out how much you eat in a week?
  How much do you eat in a whole year?

ICT
• Ask an adult or a friend to make a document and type in
  the names of the plants and animals in a food chain.
  The names should be all jumbled up. Can you put the
  food chain into the correct order, with the top predator
  at the top of the chain and the green plant at the
  bottom?

• Which foods do the people in your class like best? Do a
  survey to find out everyone's favourite foods. Put the
  information into a database and find out which is the
  favourite food overall.

Glossary

algae Organisms containing chlorophyll that live in damp or wet places.

bacteria Organisms made up of a single cell, which can only be seen through a microscope.

chemical reactions Changes that take place when chemicals are mixed together or broken down.

consumers Animals that feed on plants, or on other animals.

Equator An imaginary line around the middle of the Earth.

habitat The natural home of any plant or animal.

invertebrates Animals that do not have backbones, such as worms.

liverworts A group of green flowerless plants that live in wet places.

native Belonging to a particular place, instead of being transported there.

organisms Any living thing.

oxygen A gas that all animals and plants need to survive.

pigment A coloured substance.

predator An animal that eats other animals.

prey An animal that is eaten by another animal.

producers Plants that store the sun's energy.

rain forests Thick forests that grow where the weather is wet all year round.

reproduce To produce new plants or animals.

scavengers Animals that feed on dead organisms.

silage An animal food produced from crops such as grass, cut while it is still green and juicy.

Further Information

BOOKS

Survival: Could You Be an Otter? by R. Tabor (Heinemann Educational, 1997)

Predators in the Rainforest by Saviour Pirotta (Wayland, 1998)

Straightforward Science: Food Chains by Peter Riley (Watts, 1998)

The Hunt for Food by Anita Ganeri (Belitha Press, 1994)

WEBSITES

www.foe.co.uk Friends of the Earth

www.ffi.org.uk Flora and Fauna International

www.gn.apc.org/pesticidetrust The Pesticide Trust

www.soilassoc@gn.apc.org.uk Soil Association

www.wwf-uk.org World Wide Fund for Nature

Index

Page numbers printed in **bold** mean that there is information about this topic in a photograph, diagram or caption.